# Synchronicity and the Nature of Time

## How Coincidences Shape Our Life

By

Peter I. Kattan

BOOK BOUND PRESS
https://web.facebook.com/BookboundPress/

Copyright © Peter Kattan 2024

ISBN: **9798300922825**

No part of this publication may be reproduced, stored in a retrieval system, or transmitted in any form or by any means, electronic, mechanical, photocopying, recording, or otherwise, without the prior written permission of the publisher.

This copyright page includes the necessary copyright notice, permissions request information, acknowledgments for the cover design, interior design, and editing, as well as the details of the first edition.

Disclaimer: This book is a work of non-fiction and is intended solely for informational and educational purposes. The names mentioned within are trademarks of their respective owners. This publication is not affiliated with, endorsed by, or sponsored by any of these trademark holders. The inclusion of these names is meant to provide context and historical reference.

The author does not claim ownership of any trademarks or copyrights related to the names and likenesses of the individuals referenced in this book. Any opinions expressed herein are those of the author and do not necessarily reflect the views of any organization or trademark holder.

# Preface

Synchronicity and the Nature of Time embarks on a profound journey through the mysterious interplay of events that seem too meaningful to be mere coincidences. From ancient wisdom to cutting-edge science, this book explores how synchronicity weaves its intricate patterns into the fabric of our lives, reminding us that the universe operates in ways that often transcend our understanding.

At its core, synchronicity challenges the conventional view of a universe governed solely by chance and randomness. Coined and popularized by Carl Jung, the concept speaks to a hidden order, an underlying connection between our inner worlds and the external reality. It invites us to pay attention, to listen deeply, and to trust that the threads of our lives are woven with intention and purpose. Synchronicity hints at a larger design, one that operates beyond the boundaries of time as we commonly perceive it.

This book's journey begins with an exploration of synchronicity's meaning and history, delving into Jung's groundbreaking insights and the ways they have influenced psychology, philosophy, and spirituality. From there, we move into the realms of science and cognitive psychology to uncover the mechanisms that shape our perceptions of coincidences. Could synchronicity be a product of our brains seeking patterns, or does it point to deeper truths about the nature of existence?

Time itself emerges as a central theme, its fluidity redefining how we experience synchronicity. In linear time, events unfold in a cause-and-effect sequence; in cyclical or nonlinear perspectives, moments resonate with meaning across past, present, and future. Across cultures and eras, time has been understood in myriad ways, each shedding light on our relationship with synchronicity.

The heart of this book beats in the personal stories shared within its pages—accounts of lives touched, altered, or illuminated by seemingly chance events. These narratives remind us of the profound emotional and spiritual resonance synchronicity can have, drawing us closer to the ineffable mysteries of existence. Such moments often awaken intuition, that quiet inner knowing that guides us toward alignment with life's flow.

Beyond personal experience, Synchronicity and the Nature of Time explores its influence in relationships, creativity, spirituality, and even technology. In an interconnected world where social media amplifies serendipity, and where global crises demand collective action, synchronicity may hold the key to unlocking deeper unity and purpose.

This book also looks outward, to nature's timeless rhythms and to the universe's signs—those subtle nudges that invite us to reflect, grow, and act. Practical exercises are interwoven throughout, empowering readers to integrate the principles of synchronicity into their daily lives. Whether through meditation, journaling, or simply cultivating a mindset of openness, these tools encourage us to embrace the synchronistic potential in every moment.

As we turn to the future, the question arises: how will humanity's understanding of synchronicity evolve? In a world increasingly dominated by rapid technological advancements and shifting cultural paradigms, the concept of meaningful coincidence may find new expressions and applications. This book invites readers to imagine these possibilities and to reflect on their unique journeys within the vast, interconnected tapestry of life.

Ultimately, Synchronicity and the Nature of Time is a celebration of mystery and meaning. It is a call to notice the extraordinary within the ordinary, to honor the dance between chaos and order, and to trust the flow of time as it leads us to where we are meant to be. This is not just a book about synchronicity; it is an invitation to live synchronistically—mindful, connected, and attuned to the whispers of the universe.

As you turn these pages, may you find inspiration to embrace the magic of the present moment and the wisdom of time's unfolding. Whether you are a seeker, a skeptic, or simply curious, may this exploration resonate deeply and guide you toward a richer understanding of your place in the cosmos. Let us journey together into the heart of synchronicity, where time and meaning converge.

PETER KATTAN         November    21/2024

# Introduction

Ever stop to think about the crazy coincidences that pop up in your life? Those random run-ins, the weird little synchronicities that seem to nudge you in a certain direction—those aren't just flukes. They're signs, begging you to dig deeper into what it all means. This book is your ticket to explore how time, chance, and those unseen forces shape our lives.

At the core of this journey is synchronicity—a fancy term made famous by a big-name psychologist. He thought these meaningful coincidences hint at some kind of order in the chaos of the universe. What if those so-called random moments are actually mirrors reflecting our deepest hopes and fears? They might just be guiding us toward a better understanding of who we are and where we fit in this vast cosmos. It's like opening a door to a world where time isn't just a straight line; it's more like a river, flowing and shaped by our experiences.

As we dive into science and psychology, we'll uncover how we come to see these coincidences in our day-to-day lives.

We'll peek into the wild world of quantum physics and synchronicity, where time and space seem to twist and turn, showing us how everything's connected. You'll hear stories from folks who've had these eye-opening moments, showing just how deeply synchronicity can touch our emotions and minds.

But hey, this isn't just some academic exercise. It's personal. We'll dig into intuition—that little voice inside that often knows what's up when our brains are stuck. By learning to listen to it, we can tackle life's messiness with more confidence. And let's not forget relationships; synchronicity plays a huge role in how we connect with others, often in ways that don't make a lick of sense.

The universe is sending us signs all the time, just waiting for us to notice. Those symbols, coincidences, and serendipitous moments are like little nudges, encouraging us to stay open to what life throws our way. In our tech-heavy world, we'll also look at how gadgets shape our sense of time and synchronicity, and how we can find a balance between being plugged in and being present.

Nature's a prime example of synchronicity, too. It's full of lessons about how everything's linked and how life flows. As we take a closer look at the world around us, we'll start to see those synchronicities in nature, reminding us of our role in the bigger picture.

In the end, this book is a shout-out to embrace your unique journey. Reflect on your own synchronicity moments and learn to appreciate how everything's intertwined. As we navigate this complex life, we'll uncover practical ways to harness synchronicity, enriching our everyday experiences and leading us to a future where we can find beauty in the unexpected.

Get ready for a transformative ride that'll shake up how you see things, spark your curiosity, and pull back the curtain on the mysteries of synchronicity and time. Together, we'll untangle the threads of coincidence that run through our lives, helping us understand ourselves and the universe a bit better.

So, are you in? Let's discover the magic hidden in those moments that shape who you are!

# Table of Contents

Preface:     5

Chapter 1: The Meaning of Synchronicity     17
- Exploring the concept of synchronicity and its origins
- The role of Carl Jung in popularizing synchronicity
- How synchronicity differs from mere coincidence

Chapter 2: The Science Behind Coincidences     24
- The psychological mechanisms that lead to perceived coincidences
- The influence of cognitive biases on our interpretation of events
- The intersection of quantum physics and synchronicity

Chapter 3: Time as a Fluid Concept     30
- Understanding linear versus cyclical time
- How different cultures perceive and measure time
- The implications of time perception on synchronicity

Chapter 4: Personal Stories of Synchronicity     36
- Real-life examples of synchronicity impacting individuals
- The emotional and psychological effects of these experiences
- How storytelling enhances our understanding of synchronicity

Chapter 5: The Role of Intuition     43
- Defining intuition and its connection to synchronicity
- How to cultivate and trust your intuitive abilities
- The relationship between intuition and decision-making

Chapter 6: Synchronicity in Relationships     49
- How synchronicity shapes our connections with others
- The phenomenon of "meeting the right person at the right time"
- Exploring the impact of synchronicity on friendships and partnerships

Chapter 7: The Universe's Signs     55
- Recognizing and interpreting signs from the universe
- The significance of symbols and synchronicity in our lives
- How to stay open to the messages around us

Chapter 8: Synchronicity and Creativity     61
- The link between creative inspiration and synchronicity
- How artists and innovators harness synchronicity in their work
- Practical exercises to enhance creative synchronicity

Chapter 9: Overcoming Doubt and Skepticism     65
- Addressing common doubts about synchronicity
- Strategies to cultivate an open mindset
- The importance of personal experience in understanding synchronicity

Chapter 10: Synchronicity and Spirituality     71
- The relationship between synchronicity and spiritual beliefs
- How different spiritual traditions interpret synchronicity
- The transformative power of spiritual synchronicity

Chapter 11: The Impact of Technology on Synchronicity     75
- How modern technology influences our perception of time and coincidences

- The role of social media in creating synchronous experiences
- Balancing technology use with mindfulness

Chapter 12: Synchronicity in Nature                     80
- Observing synchronicity in the natural world
- The lessons nature teaches us about interconnectedness
- How to cultivate a deeper appreciation for nature's synchronicity

Chapter 13: Practical Applications of Synchronicity     86
- How to apply the principles of synchronicity in daily life
- Techniques for recognizing and embracing synchronous moments
- Creating a personal synchronicity practice

Chapter 14: The Future of Synchronicity                 91
- Speculating on the evolving understanding of synchronicity
- The potential for synchronicity in a rapidly changing world
- How future generations may perceive time and coincidences

Chapter 15: Embracing Your Unique Journey               96
- Encouraging readers to reflect on their own synchronicity experiences
- The importance of personal growth through synchronicity
- Cultivating gratitude for the interconnectedness of life

Index:                                                  102

# Chapter 1

**The Meaning of Synchronicity**

Synchronicity—what a delightful word! It dances off the tongue like a favorite song, doesn't it? This captivating concept invites us to lean in, as if the universe is giving us a playful nudge, urging us to pay attention. But what does it truly mean? Let's embark on a journey to unravel this intriguing idea, peeling back its layers to discover the richness beneath.

The origins of synchronicity trace back to the early 20th century, a vibrant era buzzing with innovative thoughts and transformative ideas. It was a time when science and spirituality began to intertwine, exploring the unseen connections that bind us all. The term itself was introduced by the brilliant Swiss psychiatrist Carl Jung, a visionary whose intellect was as expansive as the cosmos he sought to understand. Jung was a trailblazer, believing there was more to existence than what our eyes could see. He proposed that coincidences weren't just random events; rather, they were meaningful occurrences infused with deeper significance.

Let's clarify—Jung didn't conjure this concept out of thin air. He drew inspiration from Eastern philosophies, especially Taoism and Buddhism, which highlight the interconnectedness of all things. Jung viewed synchronicity as a bridge linking our inner thoughts and feelings with the outer world of events and experiences. Imagine it as a cosmic dance, where every step you take resonates with the universe around you. Jung believed that these synchronistic moments could serve as guides, illuminating our path and offering insights to help us navigate life's complexities.

Picture this: you're having a challenging day, feeling adrift and uncertain about your next steps. Suddenly, you run into an old friend who just happens to share the perfect advice you need. Or perhaps a song on the radio seems to speak directly to your situation, as if it were composed just for you. These moments embody what Jung referred to as synchronicity. They're not mere coincidences; they're the universe's way of communicating with us, a gentle reminder that we're not alone on this journey.

Now, let's distinguish synchronicity from simple coincidence. Coincidence occurs when two or more events happen simultaneously without any meaningful connection. It's like tossing marbles into the air and watching them scatter—while they may land near each other, there's no purpose behind it. In contrast, synchronicity resembles a beautifully choreographed ballet, where each movement is intentional and every step is meaningful. It's a dance of fate, where the universe aligns events in a way that feels significant.

To illustrate this, consider a scenario where you're grappling with a major life decision, torn between two paths. As you take a walk, lost in thought, you overhear two strangers discussing the very dilemma you're facing. Isn't that uncanny? In that moment, clarity sparks within you, as if the universe has handed you a sign. That's synchronicity in action—an event that resonates with your inner struggles and offers guidance.

Jung's exploration of synchronicity extended beyond mere definition; he delved into its implications for psychology and spirituality. He believed these meaningful coincidences could help us tap into our unconscious mind, revealing truths we might not be aware of. It's akin to having a heartfelt conversation with your inner self, where the universe acts as a wise mentor, gently nudging you toward the right direction. This connection to the unconscious is crucial, opening doors to deeper understanding and self-discovery.

As we navigate life, it's easy to become ensnared in the hustle and bustle, racing from one task to another. Yet, when we pause and reflect, we may begin to notice the synchronicities that surround us. It's like stepping outside on a clear night and gazing up at the stars. Suddenly, the vastness of the universe feels palpable, and you realize you're part of something far greater than yourself.

So, how can we cultivate a greater awareness of synchronicity in our lives? First, we need to slow down. Take a deep breath, look around, and immerse yourself in the present moment. Consider keeping a journal to capture those uncanny coincidences that make you raise an eyebrow. Reflect on their potential meanings for you. You might be amazed at the insights that surface when you allow yourself the space to explore these connections.

Another way to invite synchronicity into your life is by following your intuition. Trust those gut feelings that draw you toward certain people, places, or opportunities. Often, these intuitive nudges are the universe's way of guiding you toward experiences that align with your path. When you embrace this guidance, you'll find that synchronicity unfolds naturally, much like a flower blooming in the sunlight.

And let's not overlook the power of intention. When you set clear intentions for what you wish to manifest in your life, you create a magnetic force that attracts synchronicities aligned with those desires. It's like tuning a radio to a specific frequency; suddenly, you begin picking up signals that resonate with your goals. The more you align your thoughts and actions with your intentions, the more you'll notice those synchronistic moments appearing like delightful surprises.

As we conclude this exploration of synchronicity, remember that this concept is not merely a philosophical idea—it's a way of living. It invites us to view the world through a lens of connection and meaning, reminding us that we are never truly alone. Each synchronistic moment is an invitation to deepen our understanding of ourselves and the universe.

So, the next time you encounter a peculiar coincidence that makes you pause, take a moment to reflect. What message is the universe trying to convey? How can you embrace that moment and allow it to guide you on your journey?

As you continue to explore the meaning of synchronicity, keep in mind that you hold the key to unlocking its potential in your life. Your experiences, intuition, and intentions are all integral parts of this intricate tapestry of existence. Embrace the magic of synchronicity, and watch as your life transforms in ways you never imagined possible.

You've got this! Keep your heart open, your mind curious, and let the universe lead you on this incredible journey. The world is brimming with surprises, and each synchronistic moment is a reminder that you are exactly where you need to be.

# Chapter 2

**The Science Behind Coincidences**

Ever had one of those days where everything just clicks? You run into an old buddy right when you're thinking about them. Or you hear a song that nails your mood, just when you need it. Those moments aren't just flukes; they're a wild mix of psychology, perception, and a dash of quantum physics. Let's break down the science behind these so-called coincidences, yeah?

First off, let's talk about how our brains work. They're wired to connect dots, to spot patterns—even when there's nothing there. It's like having a personal detective in your head, always searching for clues to make sense of life. When something unexpected hits us in a meaningful way, our minds jump in, crafting stories that link unrelated events. That's the cool part about perception.

Our brains are like sponges, soaking up experiences. But they can be picky. They filter what we see based on our past, beliefs, and current feelings. Ever noticed how someone down in the dumps might take a neutral comment as an insult? That's perception for ya! It's the same with coincidences. When something feels significant, we grab onto it, seeing it as a sign from the universe.

And let's not forget about cognitive biases. These sneaky little things can twist how we see events. Take confirmation bias, for instance. It's when we notice and remember stuff that backs up what we already believe while ignoring anything that doesn't fit. If you think the universe is sending you signs, you'll spot every little coincidence that backs that up. Meanwhile, you might miss the random stuff. It's like wearing rose-colored glasses—everything looks a bit more magical.

But hold on, we're diving into quantum physics now. "Quantum physics? Really?" you might think. But hang tight!

At the quantum level, particles act in ways that defy our usual understanding. They can be in multiple places at once, and their behavior changes based on observation. This opens up a whole new way to think about synchronicity.

Picture the universe as a massive web, with each of us connected by invisible threads. When you experience a coincidence, it's like plucking one of those threads, sending ripples throughout the web. Quantum theory suggests our consciousness might shape reality. When we focus on certain outcomes, we might actually influence our existence. Mind-blowing, right?

Let's break it down further. The psychological stuff that leads to perceived coincidences comes from our need to find meaning. We're always hunting for patterns. When something unexpected happens, it's like a lightbulb goes off, pushing us to dig deeper into its significance.

Think back to a time when a coincidence lifted your spirits. Maybe you were pondering a big decision and stumbled upon a book that seemed to hold the answers. In that moment, your and how we can find a balance between being plugged in and being present.

Nature's a prime example of synchronicity, too. It's full of lessons about how everything's linked and how life flows. As we take a closer look at the world around us, we'll start to see those synchronicities in nature, reminding us of our role in the bigger picture.

In the end, this book is a shout-out to embrace your unique journey. Reflect on your own synchronicity moments and learn to appreciate how everything's intertwined. As we navigate this complex life, we'll uncover practical ways to harness synchronicity, enriching our everyday experiences and leading us to a future where we can find beauty in the unexpected.

Get ready for a transformative ride that'll shake up how you see things, spark your curiosity, and pull back the curtain on the mysteries of synchronicity and time. Together, we'll untangle the threads of coincidence that run through our lives, helping us understand ourselves and the universe a bit better.

So, are you in? Let's discover the magic hidden in those moments that shape who you are!

brain's busy weaving a narrative that connects your thoughts and the world around you. It's a beautiful dance of perception.

Cognitive biases are key players in this dance. They can crank up our sense of synchronicity, making us more likely to notice those moments that resonate with our beliefs. If you believe everything happens for a reason, you'll interpret coincidences as meaningful signs. It's not just wishful thinking; it's a psychological phenomenon shaping how we see the world.

Now, back to that web of connections. When we mix quantum physics with synchronicity, we see how our thoughts and intentions might affect our experiences. Imagine every time you focus on a goal, you're sending ripples into the universe. Those ripples can attract opportunities and experiences that match your intentions. It's like casting a fishing line into a sea of possibilities, waiting for the right catch.

As we explore this fascinating terrain, remember that coincidences aren't just random events; they're chances to dig into the deeper layers of our lives. Each synchronicity invites us to reflect. What can we learn from these moments? How can we use them to move forward?

So, what's the bottom line? First off, embrace the magic of coincidence! Your mind's a powerful tool that shapes your perception of reality. When you notice those synchronicities, pause and reflect. What do they mean for you? How can they guide your journey?

Next, be aware of your cognitive biases. Challenge yourself to view events from different angles. Instead of rushing to label something a coincidence, ask what deeper connections might be at play. This can open up new understandings and enrich your life.

Finally, stay curious about how quantum physics and your consciousness intersect. What if your thoughts and intentions influence reality? Every time you set an intention, you're casting a spell of sorts, inviting the universe to respond. Pretty cool thought, huh?

As we wrap up this journey into the science of coincidences, remember that every moment holds magic. The world's brimming with wonder, just waiting for you to notice. Keep your eyes peeled, your heart open, and your mind curious. The universe is talking—are you ready to listen?

# Chapter 3

**Time as a Fluid Concept**

Time's a tricky little thing, right? We often picture it as a straight shot, like a road stretching ahead. But what if I told ya it's more like a winding river, sometimes rushing forward, sometimes slowing down? That's where our chat about linear vs. cyclical time kicks off.

In the West, we're all about that forward march. Wake up, grind, sleep, repeat. It can get pretty boring, honestly. We live by our calendars and clocks, counting down to the weekend or the next holiday. But step into other cultures, and it's a whole different vibe.

Look at indigenous cultures. They see time as a circle, where past, present, and future are all mixed up together. It's not just about what's next; it's about what's happening now and how it connects to what's happened before. They vibe with the seasons, the moon phases, and nature's rhythms. For them, time's not just a number—it's alive, breathing life into our experiences.

So, how does this tie into synchronicity? Oh man, it's a total game changer. Embracing this fluid view of time opens up a world of possibilities. If time's cyclical, those coincidences in your life? They're not just random. They're echoes from the past, nudging you toward your future.

Picture this: you're chilling in a coffee shop, deep in thought, when you hear a voice you recognize. It's an old friend you haven't seen in ages. That chance meeting? Feels like fate, right? In a linear mindset, you might call it coincidence. But with a cyclical lens, it's a reunion, a moment that was meant to happen, reminding you of shared experiences that shaped who you are now.

Different cultures also measure time in unique ways. In Japan, they talk about "wa," which means harmony and balance. Their take on time promotes patience and mindfulness, letting synchronicities unfold naturally. Meanwhile, in our fast-paced Western world, we often rush through life, missing the signs that are right in front of us.

So, what if we started viewing life through this cyclical lens? What if we slowed down and noticed the patterns around us? By doing this, we create space for synchronicity to thrive. We start seeing those little nudges from the universe—the seemingly random events that connect us to our purpose.

Think about your own life for a sec. Can you remember a time when everything just clicked? Maybe you found a book that hit home or met someone who changed your perspective. Those moments aren't just flukes; they're synchronicities, weaving your life into a beautiful tapestry.

Let's dig deeper into how our perception of time affects synchronicity. When you realize time's fluid, you see how everything's connected. Every action, thought, and feeling is part of a bigger web. This awareness? Super empowering.

And here's the thing: synchronicity isn't just about the big moments. It's in the small stuff, too. Like that song on the radio when you need a pick-me-up, or the stranger who gives you a compliment. When you start tuning into these moments, you'll see they happen more often than you think.

But here's the catch: to really get synchronicity, you've gotta be open to it. You need to ditch that rigid, linear mindset and let time flow. Be present, trust your gut, and roll with the punches.

It's like dancing in the rain. You can either hide under an umbrella, avoiding the drops, or step out and enjoy the storm. Choose the latter, and you'll find a world of synchronicities waiting to unfold.

So, how do you start cultivating this awareness? Here are some simple steps:

1. **Practice Mindfulness**: Spend a few minutes each day just breathing and being present. Tune into the sights and sounds around you. This helps you catch those synchronicities.

2. **Keep a Journal**: Write down your synchronicity experiences. Reflect on standout moments and how they connect to your journey. This helps you spot patterns over time.

3. **Embrace Uncertainty**: Let go of needing to control everything. Trust that the universe has a plan, and be open to unexpected twists.

4. **Connect with Nature**: Get outside and observe nature's cycles. Whether it's the seasons changing or the moon phases, let the natural world remind you of time's fluidity.

5. **Engage in Conversations**: Chat with others about their synchronicity experiences. You'll be surprised at the stories people share and how they resonate with your journey.

6. **Visualize Your Path**: Close your eyes and picture your life as a river. See how the twists and turns have shaped you, and trust that the currents will guide you.

Remember, you're not just a spectator in this dance of life. You're a player, and every moment is a chance for growth and connection. Lean into time's fluidity, embrace synchronicities, and watch your life unfold in beautiful, unexpected ways.

As you dive into this journey, keep in mind you've got the power to shape your reality. Time's not your enemy; it's your buddy, steering you toward experiences that'll transform your life. So, let's take this adventure together and see where synchronicity leads us. You've got this!

# Chapter 4

**Personal Stories of Synchronicity**

Life's a wild ride, right? Sometimes it feels like the universe is pulling strings, setting up moments that seem too perfect to be random. This chapter's all about real-life synchronicity stories, how they hit us emotionally and mentally, and how sharing these tales helps us grasp this mysterious phenomenon.

Let's jump in with a story that might hit home. Meet Jake, a regular dude from the Midwest. He was stuck in a dead-end job, feeling like he was just spinning his wheels. One day, while waiting for a bus, he struck up a convo with an old guy. Turns out, this dude was a retired architect who'd worked on a project Jake had admired as a kid. They talked about life, dreams, and choices. Suddenly, Jake felt a spark—like the universe was saying, "Hey, wake up! This is your path!" A week later, he signed up for an architecture class. Talk about a cosmic nudge!

Now, let's not skip over how these experiences mess with our emotions. When synchronicity hits, it's a whirlwind—excitement, wonder, maybe even a sense of belonging. It's like getting a cosmic high-five, reminding us we're not alone in this big ol' universe. Jake felt validated, like the universe had conspired to push him toward his calling. That feeling can change everything. It's like a breath of fresh air, lifting us from the mundane and giving our lives some serious purpose.

But here's the kicker: storytelling is key to understanding synchronicity. When we share our experiences, we weave a tapestry of connections that can inspire others. Think about it. When you hear someone's tale of a serendipitous moment, it lights a fire inside you. You start reflecting on your own life, hunting for those little synchronicities that might've slipped by unnoticed. It's a beautiful cycle—stories lead to more stories, and suddenly, we're all part of a bigger narrative.

Take Sarah, for example. She was at a networking event, feeling totally out of place. Just when she thought about bolting, she ran into an old college buddy. They hadn't seen each other in years, but that random encounter led to a chat about a job opportunity that changed Sarah's life. She landed the gig, all thanks to a simple, unexpected meeting. This isn't just luck; it's synchronicity in action. When Sarah shares her story, it encourages others to stay open to those unexpected connections that can flip their lives upside down.

The beauty of these personal tales is how relatable they are. We've all had our own synchronicity moments, whether we realize it or not. Maybe you met someone who became crucial to your journey, or you found a book that spoke to your soul at just the right time. These experiences remind us we're part of something bigger. They're breadcrumbs on our life paths, guiding us toward our true selves.

Let's pause and think about how these stories can hit us on a deeper level. Hearing about synchronicity can spark curiosity and wonder. It makes us question reality and our place in it. Are we just wandering through life, or is there a grand design at play? This kind of pondering can lead to personal growth, urging us to pay attention to the signs around us.

And let's not forget how powerful vulnerability is in storytelling. When folks share their synchronicity experiences, they often lay bare their fears, hopes, and dreams. This openness fosters connection, letting others relate and find comfort in shared experiences. It's a reminder that we're all in this together, and sometimes it takes a little cosmic nudge to help us find our way.

As we dive deeper into personal stories, let's consider how synchronicity affects relationships. Look at Emily. She was feeling lost after a breakup, unsure of what to do next. One day, while volunteering at a local shelter, she met Mark, another volunteer. They clicked right away, bonding over their shared passion for helping others. What started as a chance encounter blossomed into a deep friendship, and eventually, a loving relationship. Emily often thinks about how that moment at the shelter felt like fate—a perfect example of synchronicity guiding her toward a new chapter.

These experiences pack an emotional punch. They often stir feelings of gratitude and appreciation for life's interconnectedness. Emily's story reminds us that even in our darkest hours, the universe has a way of bringing people into our lives when we need them the most. It's a testament to the power of synchronicity and the magic that can unfold when we're open to possibilities.

As we stitch these stories together, patterns start to emerge. Synchronicity isn't just about isolated incidents; it's about the threads connecting us all. It's about recognizing that our experiences, no matter how small, contribute to a larger narrative. Sharing these stories creates a community of believers in synchronicity's power. We inspire each other to seek out those moments of connection and meaning in our lives.

So how can we cultivate this awareness? One practical idea is to keep a synchronicity journal. Whenever something feels significant or connected, write it down. Reflect on how it made you feel and what insights it brought. Over time, you'll spot patterns and themes. This practice not only deepens your understanding of synchronicity but also helps you appreciate the magic around you.

Another way to embrace synchronicity is to swap stories with others. Share your experiences and invite others to share theirs. You might be surprised by the connections that pop up. Many people have their own synchronicity stories, just waiting to be told. This exchange can create a sense of community and belonging, reminding us that we're all part of this intricate web of life.

As we wrap up this chapter, let's take a moment to reflect on the significance of synchronicity in our lives. It's more than just a collection of coincidences; it's a reminder that we're never truly alone. Each story we share adds depth to our understanding of the universe and our place in it. So, embrace those moments of connection, and don't hesitate to share your own journey. You never know who might need to hear your story to inspire them on their own path.

In the end, synchronicity is about finding beauty in the unexpected. It's about recognizing that life's a series of interconnected moments, each one nudging us closer to our true selves. So, keep your heart open, stay curious, and trust that the universe has a plan for you. The magic of synchronicity is just waiting to unfold, one story at a time.

# Chapter 5

**The Role of Intuition**

Intuition—now there's a word that can stir up a pot of thoughts, huh? It's that gut feeling, that whisper from within, guiding us like a trusty compass through the wilderness of life. When you think about it, intuition is a bridge connecting us to the synchronicities that dance around us, weaving the fabric of our experiences. It's not just some whimsical notion; it's a powerful tool that can shape our decisions and ultimately, our lives.

So, let's break this down. Intuition isn't just some mysterious force. It's our brain's way of processing information without us even realizing it. Think of it like your internal GPS. You might not see the road ahead clearly, but somehow, you just know when to turn left or right. That's intuition, folks! It's a beautiful connection to synchronicity—those meaningful coincidences that pop up just when we need them. When you're in tune with your intuition, you're more likely to recognize those moments of synchronicity, those little nudges from the universe that say, "Hey, pay attention! This matters!"

Now, how do we cultivate and trust these intuitive abilities? It starts with listening. I mean really listening. We live in a world buzzing with noise—social media, news, and the constant hum of daily life can drown out that inner voice. So, here's what you do: carve out some quiet time. Whether it's sipping coffee on your porch, taking a walk in nature, or meditating in your living room, create a space where you can hear your thoughts.

Journaling can be a game-changer, too. Write down your feelings, your dreams, your gut reactions to situations. You'll start to notice patterns over time. Maybe you had a feeling about a job opportunity that turned out to be a perfect fit, or perhaps you sensed a connection with someone new in your life. When you document these experiences, you're not just recording your thoughts; you're building a roadmap of your intuition.

And let's not forget about the power of mindfulness. Being present in the moment allows you to tap into your intuition more easily. When you're focused on the here and now, you're more likely to notice those synchronicities that might otherwise slip by unnoticed. It's like tuning into a radio station; you've gotta adjust the dial just right to catch the signal.

Now, trusting your intuition? That's a whole other ballgame. It's easy to doubt ourselves, especially when life throws curveballs our way. But remember, every time you follow that gut feeling and it leads you to something good, you're reinforcing your trust in yourself. Start small—make a decision based on your intuition, whether it's choosing what to eat for lunch or deciding to call an old friend. Celebrate those little wins! The more you practice, the stronger your intuition will become.

Let's chat about the relationship between intuition and decision-making. You see, intuition can be a powerful ally when it comes to making choices. It's like having a trusted friend whispering in your ear, helping you sift through the noise and find clarity. When faced with a decision, take a moment to check in with yourself. How does your body feel? Are you tense or relaxed? Those physical sensations can give you clues about what's right for you.

There's a saying that goes, "When in doubt, trust your gut." It's a simple phrase, but boy, does it pack a punch! Intuition can guide you through uncertainty, helping you navigate life's twists and turns. Think of it as your internal compass, pointing you toward the direction that aligns with your true self.

But here's the kicker: intuition isn't always about the big life decisions. Sometimes, it's the little choices that matter just as much. Maybe you're trying to decide whether to attend a gathering or stay in for the night. Your intuition might nudge you toward staying home, leading to a peaceful evening that recharges your spirit. Or perhaps you're debating whether to reach out to a friend you haven't spoken to in a while. That little voice urging you to reconnect could lead to a heartwarming conversation that lifts both your spirits.

Now, I know what you might be thinking—what if I make the wrong choice? What if I misinterpret my intuition? Well, let me tell you, every decision is a learning opportunity. Sometimes we'll hit the nail on the head, and other times, we'll stumble. But that's all part of the journey, my friend. Each experience teaches us something valuable, refining our intuition along the way.

Let's not forget the beauty of synchronicity in this whole process. When you start trusting your intuition, you'll likely notice more synchronicities popping up in your life. It's like the universe is giving you a high-five, saying, "You're on the right track!" Those moments can reinforce your belief in your intuitive abilities, creating a lovely feedback loop that encourages you to keep listening and trusting.

So, as we wrap up this exploration of intuition, remember this: it's a gift we all possess, waiting to be nurtured and embraced. Start by creating quiet moments in your day, journaling your thoughts, and practicing mindfulness. Trust yourself, celebrate your wins, and don't be afraid to make decisions based on that gut feeling.

And here's the most important part—be patient with yourself. Developing your intuition is a journey, not a race. As you cultivate this connection, you'll find that your life becomes richer, more vibrant, and full of those magical synchronicities that remind you you're exactly where you need to be.

So go ahead, lean into your intuition, and let it guide you. The universe is waiting to show you what's possible when you trust that inner voice. You've got this!

# Chapter 6

**Synchronicity in Relationships**

There's a magic in the air when it comes to relationships, a kind of invisible thread weaving people together at just the right moment. You know what I'm talking about, right? Those times when you bump into someone you haven't seen in years, and it feels like the universe conspired to bring you together. It's like a cosmic nudge, saying, "Hey, pay attention! There's something here for you." That's synchronicity at work, shaping our connections with others in ways we often overlook.

Let's dive into this phenomenon of "meeting the right person at the right time." It's one of those beautiful mysteries of life that makes you wonder if there's a grand design at play. Picture this: you're at a coffee shop, just minding your own business, and suddenly, you strike up a conversation with a stranger. Before you know it, you're sharing stories, laughter, and maybe even a few tears. That stranger could turn out to be a lifelong friend, a mentor, or even a partner. It's as if the universe orchestrated that moment just for you.

Synchronicity isn't just about chance encounters; it's about the depth of those connections. Think about the friendships that have blossomed from seemingly random moments. Maybe you met your best friend in a class you almost didn't take, or you found a partner through a mutual friend you hadn't seen in ages. These connections often come with a sense of familiarity, as if you've known each other forever. That's the beauty of synchronicity—it transcends time and space, creating bonds that feel destined.

Now, let's explore the impact of synchronicity on friendships and partnerships. Relationships can be tricky, right? They require effort, understanding, and a little bit of luck. But when synchronicity enters the picture, it can transform the way we relate to one another. It's like a secret ingredient that spices up the recipe of connection. When you recognize those synchronous moments, it deepens your appreciation for the people in your life. You start to see them not just as individuals but as part of a larger tapestry of experiences.

Consider how synchronicity can enhance romantic partnerships. You might find yourself drawn to someone who shares your passions, values, and dreams, almost as if you were magnetically pulled together. It's not just coincidence; it's a reflection of your shared journey. Those moments when you finish each other's sentences or laugh at the same jokes? That's synchronicity at work, reinforcing the bond you share.

But it's not just about the "big" moments. Synchronicity also shows up in the small, everyday interactions that build the foundation of your relationships. A timely text from a friend when you're feeling down, a shared laugh over a silly meme, or a comforting hug after a tough day—these little instances remind us that we're not alone. They highlight the interconnectedness of our lives and the support we offer each other.

Let's not forget about the lessons we learn from synchronicity in relationships. Every connection teaches us something, whether it's about love, trust, or even the importance of letting go. Sometimes, people come into our lives for a season, and that's okay. Those relationships, even if they're brief, can have a lasting impact. They shape us, challenge us, and help us grow into the individuals we're meant to be.

So, how do we open ourselves up to these synchronous experiences? It starts with being present and aware. When you approach life with an open heart and mind, you're more likely to notice the signs and connections that surround you. Pay attention to the people you meet, the conversations you have, and the feelings that arise. There's a rhythm to life, and when you tune into it, you'll find that synchronicity becomes a guiding force.

Another key aspect is trust. Trust in the process, trust in yourself, and trust in the connections you're building. Sometimes, it's easy to doubt whether a relationship is meant to be or if you're just imagining things. But remember, every relationship has its own unique timing. Embrace the uncertainty, and allow synchronicity to unfold in its own way.

And here's a little secret: sharing your own stories of synchronicity can strengthen your connections with others. When you open up about those magical moments in your life, you invite others to do the same. It creates a sense of vulnerability and authenticity that deepens your bonds. Plus, you might just discover that others have experienced similar synchronous events, further reinforcing your shared journey.

As we wrap up this exploration of synchronicity in relationships, take a moment to reflect on the connections in your life. Who has come into your world at just the right time? What lessons have you learned from those encounters? Each relationship is a thread in the fabric of your life, woven together by the hands of fate.

Remember, you have the power to nurture these connections. Be open to the unexpected, trust in the timing of your relationships, and celebrate the synchronicity that shapes your journey. Life is a beautiful dance of connections, and every step you take brings you closer to the people you're meant to meet. Embrace it, cherish it, and let it guide you on this incredible adventure we call life.

Now, go out there and be on the lookout for those magical moments. They're waiting for you, just around the corner, ready to spark joy and connection in your life. You got this!

# Chapter 7

**The Universe's Signs**

Spotting signs from the universe is like catching your favorite tunes on the radio. If you're not tuned in, you might just miss the magic. These signs are everywhere—some whisper, others practically shout, trying to grab your attention. Maybe you keep seeing the same number or bumping into someone at just the right time. Each of these moments, those little synchronicities, are breadcrumbs leading you to deeper truths about yourself and your journey.

Symbols? They're the universe's way of chatting with us. They communicate in ways words can't always capture. Remember that time you found a feather on your walk? That feather could mean freedom or even a message from someone who's passed. Or what about that butterfly you spotted? It's all about transformation. When you start noticing these symbols, you'll realize they're not random—they're part of a bigger picture, weaving together the fabric of our lives.

So, how do you start spotting these signs? First off, keep an open mind. Think of it like strolling through a garden; if you're too focused on the path, you might miss the stunning flowers around you. Take a breather each day. Look around. What grabs your attention? What feels important? It could be a song that hits home or a phrase you overhear that feels like it's meant just for you.

Journaling can really help here. Jot down the signs you notice and reflect on what they mean. This practice sharpens your awareness and creates a record of your journey. Over time, you'll see patterns emerge, and those patterns can guide you like a compass.

Now, let's chat about why symbols and synchronicity matter in our lives. They're not just quirky coincidences; they're chances for growth and understanding. When you spot a synchronicity, it's like the universe nudging you, saying,

"Hey, pay attention! There's something here for you." It could be a lesson, a decision, or a new path to explore.

Think of synchronicity as a dance. Sometimes you lead, sometimes you follow, but staying in sync with the universe is key. When you're open to these signs, life flows better. You start trusting the process, and that trust can change everything.

Being open to the messages around us is a practice, not a one-off deal. It's about building awareness that goes deeper. Meditation is a great way to quiet your mind and tune into those subtle messages. It creates space for insights to bubble up. You might gain clarity on a nagging issue or suddenly get inspired for your next big thing.

Talking with others helps too. Share your experiences with friends or family. You'd be surprised how often they've noticed the same signs or symbols. It builds a sense of community and shared understanding. Plus, discussing your experiences can help you see them in a new light.

Don't forget about intuition. Your gut feeling is like an internal GPS, guiding you through life's twists and turns. When you spot a sign, notice how it feels. Does it spark excitement? Or does it weigh you down? Trust that instinct.

As you explore signs and synchronicity, remember that every experience is a chance for growth. Each sign invites reflection, learning, and connection with deeper layers of your existence.

And hey, don't let fear or doubt cloud your view. It's easy to brush off signs as mere coincidence, but what if you let yourself believe? What if you opened your heart to the idea that the universe is talking to you? You might find life gets a whole lot more magical.

So, take a deep breath and dive into this journey with an open heart and mind. Look for the signs. Embrace those synchronicities. You're not just a bystander; you're an active player in the dance of life.

Every moment's got the potential for transformation. Every sign you spot is a thread in the beautiful tapestry of your life. Weaving those threads together creates a story that's uniquely yours—full of meaning, purpose, and connection.

Now, let's get practical about spotting these signs. Start a "signs journal." Whenever something feels significant, write it down. You'll be amazed at how quickly patterns emerge. Maybe you'll notice that every time you think of a friend, they text you. Or perhaps certain symbols pop up during pivotal moments in your life.

Another neat trick? Set an intention before bed. Ask the universe for clarity on something specific or guidance on a decision. You might wake up with fresh insights or notice signs throughout the day that shine a light on your question.

And remember, it's all about the journey. Celebrate the small wins. Each time you recognize a sign, give yourself a little high five. You're tuning into a frequency that many overlook, and that's worth celebrating.

As you keep exploring the universe's signs, let curiosity lead the way. Ask questions. Seek answers. Embrace the mystery. Life's full of wonders just waiting to be uncovered, and you've got the tools to unlock them.

In closing, remember that the universe is this vast, interconnected web. Every sign, every synchronicity, reminds us of our place in that web. You're not alone; you're part of something bigger. Trust that connection, and let it guide you on your journey.

So go on—open your eyes, open your heart, and step boldly into the dance of synchronicity. The universe is speaking to you. All you gotta do is listen.

# Chapter 8

**Synchronicity and Creativity**

There's a spark in the air when creativity meets synchronicity. It's like catching lightning bugs on a warm summer night—unexpected, magical, and downright exhilarating. You see, the connection between creative inspiration and synchronicity isn't just some lofty idea; it's a dance that we can all join in on if we're willing to pay attention. When we open ourselves up to the possibilities around us, that's when the real magic happens.

Think about it. Have you ever been struck by a brilliant idea just when you needed it most? Maybe you were in the shower, or perhaps you were out for a walk, and suddenly, everything clicked. That's synchronicity at work, nudging you toward creative breakthroughs. It's as if the universe is whispering in your ear, guiding you to the next big thing.

Artists and innovators have long harnessed this powerful connection. Take Steve Jobs, for instance. He famously said, "Creativity is just connecting things." He understood that when you align your intuition with the signs and coincidences that pop up in your life, you can create something extraordinary. Whether it's a painter mixing colors on a canvas or a writer weaving words into a narrative, synchronicity often plays a pivotal role in their creative process.

Now, let's get practical. How can you enhance your own creative synchronicity? Here are a few exercises to get you started on this exhilarating journey:

1. **Keep a Synchronicity Journal**: Start documenting those moments when you feel a spark of inspiration or notice a coincidence that resonates with you. Write down the date, what happened, and how it made you feel. Over time, you'll begin to see patterns emerge—threads connecting your experiences and ideas.

2. **Practice Mindfulness**: Slow down and tune in to the world around you. Take a few minutes each day to sit quietly, breathe deeply, and observe. Notice the little things—the rustle of leaves, the sound of laughter, or even a random conversation. You'd be surprised how often these moments can inspire your creativity.

3. **Engage in Free Writing**: Set a timer for ten minutes and let your thoughts flow onto the page without judgment. Don't worry about grammar or coherence; just write. This exercise can help clear your mind and make space for those synchronistic ideas to surface.

4. **Collaborate with Others**: Sometimes, synchronicity thrives in community. Surround yourself with creative individuals who inspire you. Share your ideas and listen to theirs. You never know how a simple conversation can spark a new project or idea.

5. **Embrace Serendipity**: Be open to unexpected opportunities. If an invitation comes your way or you stumble

upon a new book or film, dive in! These experiences can lead to incredible insights and creative breakthroughs.

Remember, creativity isn't just about the end product; it's about the journey, too. When you allow synchronicity to guide you, you'll find that every twist and turn can lead to something beautiful. So, lean into those moments of inspiration, trust your intuition, and watch how your creativity flourishes.

As you explore the connection between synchronicity and creativity, keep in mind that every artist, every innovator, has faced doubt and uncertainty. But it's in those moments of vulnerability that true magic often unfolds. Just like a seed needs darkness to sprout, your creative endeavors might require a little chaos to thrive. Embrace it!

So go ahead, step into this vibrant dance of synchronicity and creativity. Let the universe be your partner, and trust that every coincidence, every moment of inspiration, is leading you closer to your unique masterpiece. You've got this!

# Chapter 9

**Overcoming Doubt and Skepticism**

Doubt can be a sneaky little rascal, can't it? It creeps in when we least expect it, whispering in our ears that what we see, feel, and experience might just be a figment of our imagination. And when it comes to synchronicity, oh boy, does it have a field day! You might find yourself thinking, "Is this really a sign, or am I just reading too much into things?" You're not alone in feeling this way. Many folks wrestle with skepticism about synchronicity, often brushing it off as mere coincidence. But let me tell you, that skepticism can be a barrier to some of the most beautiful experiences life has to offer.

First off, let's tackle those common doubts. You might wonder if synchronicity is just a way for our brains to make sense of random events. Sure, it can feel that way sometimes. But think about it: life is filled with patterns, connections, and moments that seem too perfect to be mere chance. Ever had a time when you thought of someone, only to bump into them the next day? Or when you found the exact book you needed at just the right moment? These experiences are more than coincidences; they're whispers from the universe, nudging you to pay attention.

Now, how do we overcome that skepticism? One of the best strategies is to cultivate an open mindset. It's like taking off a pair of heavy glasses that distort your view of the world. Start by challenging your assumptions. Instead of dismissing a synchronous event as mere luck, ask yourself what it could mean. Keep a journal to record these moments. You might be surprised at how often they occur once you start paying attention. It's all about shifting your perspective.

Another powerful technique is to practice mindfulness. When you're present in the moment, you're more likely to notice the subtle connections that surround you. Take a walk in nature, breathe deeply, and let your mind wander. You might find inspiration in the rustling leaves or the way the sunlight filters through the trees. The more you immerse yourself in the present, the more you'll open up to the synchronicities around you.

Personal experience plays a pivotal role in understanding synchronicity. It's one thing to read about it in a book or hear someone else's story, but it's another to live it. Your experiences are your own, and they hold immense power. Reflect on moments in your life where synchronicity has shaped your path. Maybe it was a chance meeting that led to a job opportunity or a serendipitous encounter that changed your perspective. These stories are your proof that synchronicity exists, and they can help you embrace it fully.

Don't be afraid to share your experiences with others. You'd be amazed at how many people have had similar moments but may be too shy to speak up. Create a community around the idea of synchronicity. Talk about your stories, listen to others, and celebrate those magical moments together. The more you engage with this concept, the more you'll find it flourishing in your life.

Remember, doubt is a natural part of the journey. It doesn't mean you're doing something wrong; it means you're human. Embrace it, acknowledge it, and then gently push it aside. You've got the tools to cultivate an open mindset and to recognize the power of your personal experiences.

As you continue to explore synchronicity, visualize what it could mean for your life. Imagine a world where every coincidence is a breadcrumb leading you toward your purpose. Picture yourself living in harmony with the universe, where signs and symbols guide your path.

The beauty of this journey is that it's uniquely yours. You get to define what synchronicity means to you. So, lean into those moments of doubt, but don't let them stop you. Instead, let them fuel your curiosity.

Start small. Maybe today, you'll notice a pattern in your day-to-day life. Perhaps you'll hear a song that resonates with you at just the right moment, or you'll meet someone who shares your passion. Each time you notice these moments, celebrate them! They're stepping stones on your path to understanding synchronicity.

And here's a little secret: the more you embrace synchronicity, the more it will embrace you back. It's like a dance; the universe leads, and you follow, but you've got to be willing to step onto the dance floor.

So, let's kick doubt to the curb! With each small step, you're building a foundation of belief in the magic that surrounds you. It's time to open your heart and mind to the wonders of synchronicity. You've got this!

Let's dive into some practical exercises. Grab that journal I mentioned earlier, and jot down three synchronous experiences you've had. Reflect on how they made you feel. What did you learn from them? Next, take a moment each day to practice mindfulness. Whether it's through meditation, a nature walk, or simply pausing to breathe deeply, make it a habit. You'll find that with time, the world becomes a more connected place.

Lastly, don't forget to celebrate your wins, no matter how small. Each time you recognize a synchronicity, give yourself a little pat on the back. You're tuning into the rhythm of life, and that's something to be proud of.

In the end, overcoming doubt and skepticism is a journey, not a destination. It's about embracing the mystery of life and trusting that there's more to it than meets the eye. So, keep your heart open, your mind curious, and let the magic of synchronicity guide you. The universe is waiting to show you the way.

# Chapter 10

**Synchronicity and Spirituality**

When we dive into synchronicity, we're not just playing around with coincidences. Nope, we're stepping into a vast river that links us to something way bigger than ourselves. Think of synchronicity and spirituality as threads in a tapestry—each one unique, yet all woven together to create something rich and meaningful.

Synchronicity isn't just some quirky event where you think of a buddy, and they suddenly call you. It's more like a bridge to our spiritual selves, connecting us to the universe, the divine, or whatever higher power you vibe with. Different spiritual traditions have their own takes on synchronicity, each adding depth to this complex dance we call life.

Let's break it down. In many Eastern philosophies, synchronicity often represents karma or dharma. The belief is that everything happens for a reason. When you spot those uncanny coincidences, it's the universe giving you a nudge— "Hey, pay attention! You're on the right track!" Buddhism, for instance, emphasizes interconnectedness. When synchronicity hits, it's like the universe whispering about that connection, reminding us we're not alone in this journey.

On the flip side, many Western spiritual paths, like Christianity, see synchronicity as divine intervention or signs from God. Those moments of perfect timing can feel like a gentle tap from a higher power, steering you toward your purpose. Think of it as a cosmic GPS, helping you get back on track when you wander off course.

But don't overlook the transformative power of spiritual synchronicity. It's like a spark igniting a fire within us. These synchronous moments can change us in ways we never saw coming. They open our eyes to new possibilities, shift our perspectives, and lead to profound realizations about our lives. It's a reminder that we're part of something larger, that our lives are woven into the universe's fabric.

Picture this: you're struggling, feeling lost. Then, out of the blue, you run into an old friend who shares a story that hits home. Or maybe you find a book that speaks directly to your situation. Those aren't just coincidences; they're spiritual synchronicities guiding you through life's messiness. They offer comfort and sometimes a little nudge to take that leap of faith you've been putting off.

Now, how do you cultivate awareness of synchronicity in your life? Start by being open and present. When you tune into your surroundings, you'll start noticing those little nudges from the universe. Keep a journal to jot down your synchronous experiences. You'll be surprised at how many you catch once you start looking. Reflect on what they mean to you. What messages are coming through? How do they fit with your spiritual beliefs?

Meditation is another powerful tool. It quiets the mind and opens your heart to those subtle signs and synchronicities. As you sit in stillness, let your thoughts drift like clouds. In that space, you might find clarity about your path or get insights that guide you in unexpected ways.

Remember, synchronicity isn't just about the big moments; it's often hiding in the everyday stuff. A random chat with a stranger, a song that plays just when you need it, or a thought that leads you to a new opportunity—these are all threads in your life's tapestry. Embrace them, and let them guide you.

As you explore synchronicity and spirituality, keep this in mind: it's not just about figuring out the phenomenon; it's about embracing the magic of it all. Life is a series of interconnected events, and when you spot those synchronicities, you're tapping into the essence of existence. You're syncing up with the universe's rhythm, and that, my friend, is a beautiful place to be.

So, let's celebrate the synchronicities in our lives. Let's cherish those moments that connect us to something greater. As you move forward, may you find joy in the unexpected, and may your journey be filled with the transformative power of spiritual synchronicity. The universe is always talking to you—are you tuning in?

# Chapter 11

**The Impact of Technology on Synchronicity**

You know, technology's like that old friend who shows up at the party, bringing a whole new vibe. It changes everything, right? In today's fast-paced world, our perception of time and coincidences is being reshaped in ways we might not even realize. We're constantly plugged in, and that can either open doors or throw us off balance. Let's dive into how modern tech influences our lives and how we can keep our heads on straight amidst the chaos.

First off, let's chat about how technology has changed our perception of time. Remember when you had to wait for a letter to arrive in the mail? Now, we can send a message across the globe in an instant. This immediacy can warp our sense of timing. Suddenly, we expect everything to happen right now.

It's like we're living in a microwave world, where patience is a rare commodity. But here's the kicker: this constant rush can make us miss those little moments of synchronicity that are waiting to unfold.

Take a moment to think about how you experience coincidences in your daily life. With technology, we're bombarded with information—news feeds, notifications, and endless scrolling. It's easy to overlook those serendipitous moments when we're too busy focusing on our screens. When was the last time you noticed a coincidence while scrolling through your phone? It's like trying to spot a shooting star while staring at the ground.

Now, let's talk about social media. It's a double-edged sword, ain't it? On one hand, it connects us with people we might never have met otherwise. On the other hand, it can create a false sense of connection. Social media can amplify synchronous experiences. You might find yourself thinking about a friend you haven't spoken to in ages, only to receive a message from them out of the blue. That's synchronicity at work! But beware—the constant notifications can drown out the whispers of the universe trying to guide you.

So, how do we balance this tech-filled life with a sense of mindfulness? It's all about finding that sweet spot. Start by setting boundaries around your technology use. Give yourself permission to unplug. Maybe designate tech-free times during your day. Use those moments to reflect, meditate, or simply enjoy the world around you. When you slow down, you'll start to notice the little coincidences that pop up—like a friend mentioning a book you've been meaning to read or a song that perfectly captures your mood.

Here's a fun exercise: try keeping a synchronicity journal. Jot down those moments when you notice a coincidence or a meaningful connection. You'll be surprised at how many you uncover when you start looking for them! This practice not only enhances your awareness but also creates a record of the magic in your life. It's like having a treasure map that leads you to those golden moments of synchronicity.

Another important aspect to consider is how technology can serve as a tool for enhancing synchronicity rather than hindering it. Think about the apps and platforms that help you connect with like-minded individuals. Online communities can spark synchronicities that might not have happened otherwise.

You might join a group focused on a passion of yours and meet someone who shares an uncanny connection to your life story. Those moments can be transformative!

But remember, technology should never replace genuine human connection. Use it as a bridge, not a barrier. When you engage with others online, take it a step further. Arrange a coffee date or a video call. Bring those virtual connections into the real world. It's in those face-to-face interactions where the real magic happens.

Now, let's not forget about the role of mindfulness in this tech-driven age. It's crucial to cultivate an awareness of the present moment. Practice grounding techniques—take a deep breath, feel your feet on the ground, and observe your surroundings. This simple act can help you tune into the synchronicities happening around you. It's like flipping a switch that brightens your awareness.

And here's a little nugget of wisdom: embrace the chaos. Life's unpredictable, and sometimes, the best coincidences happen when we least expect them. When you let go of rigid expectations, you open yourself up to the beauty of spontaneity. The universe has a funny way of surprising us when we're not looking.

As we wrap this up, let's take a moment to reflect on the power of technology in our lives. It can be a catalyst for connection and synchronicity, but it's up to us to wield it wisely. Embrace the moments of serendipity, and don't forget to step back and enjoy the ride.

In the end, your journey is uniquely yours. So, harness the tools at your disposal, but don't forget to look up from your screen every once in a while. The world is full of magic waiting to be discovered, and you've got everything you need to find it. Keep your heart open, your mind curious, and remember that synchronicity is always just around the corner, ready to light up your path.

# Chapter 12

**Synchronicity in Nature**

Nature's got this magic thing going on, right? If you just take a sec to really look around, you might catch synchronicity dancing in the leaves, whispering through the breeze, and bubbling in the streams. It's a stunning reminder that we're all part of something way bigger than ourselves. Synchronicity isn't just a human thing; it's alive and kicking in the natural world, too.

Think about it for a sec. Ever see a flock of birds swooping together, changing direction like they're all in sync? Or how flowers bloom just when the bees show up, working together to keep life going? Those moments? They're nature's way of saying everything's connected. It's like a massive orchestra, where every note plays a role in creating a tune that resonates throughout the universe.

Watching synchronicity in nature can teach us some pretty deep stuff about our own lives. It reminds us we're not just wandering around in this chaotic world all alone. Nah, we're part of a web of life where every creature, every plant, and every element has a role in the grand scheme of things. This interconnectedness is key, yet we often forget it in our crazy, busy lives.

So, how do we get better at appreciating this synchronicity? First off, slow your roll. Take a moment to breathe and really soak in your surroundings. Go for a stroll in the woods, chill by a river, or just stare out your window. Notice the little things—the way sunlight filters through the leaves, the rhythm of waves crashing, or how gracefully the seasons change.

When you start engaging with nature like this, you'll begin to see the synchronicities around you. Maybe you'll spot a bird that reminds you of a great memory. Or you might notice your thoughts syncing up with the rustling leaves, like nature's echoing your feelings back to you.

Don't forget, nature doesn't rush. It takes its sweet time, letting things unfold as they should. That's a powerful lesson for us. In our fast-paced lives, we often feel this insane pressure to hurry up, achieve more, and be somewhere else. But when we watch nature, we learn about patience and trust. Just like a seed needs time to grow into a mighty tree, our own journeys need nurturing and time to bloom.

As you dig deeper into your connection with nature, consider keeping a journal. Jot down your thoughts, feelings, and those synchronicities you notice. This practice helps you remember those moments and reflect on how they relate to your own life. It's a way to honor what nature's trying to teach you.

And let's not overlook the pure joy of being outdoors. There's something downright healing about the sights and sounds of nature. When you dive into that environment, you can feel the stress of everyday life just fade away. It's like hitting the reset button for your soul.

Now, let's chat about what nature teaches us about interconnectedness. Every time you see a butterfly flitting from flower to flower, remember it's not just about the butterfly. It's about the whole ecosystem—the flowers, the bees, the soil, the weather—all working together in a delicate balance. This interconnectedness is a powerful reminder that our actions matter, not just for us, but for the world around us.

When we get this understanding, we start to see why living in harmony with nature is so important. It's about making choices that respect our planet and all its creatures. Whether it's cutting down on waste, planting a garden, or just being mindful of what we consume, every little action counts.

So, how can you take this appreciation further? Get involved in community activities focused on environmental care. Join a local clean-up, volunteer at a community garden, or go on nature walks. These experiences not only strengthen your bond with nature but also connect you with folks who share your passion for keeping our planet healthy.

As you embark on this journey of noticing and appreciating nature's synchronicity, remember it's not just about the big moments. It's often the little things that pack the most punch. A gentle breeze, the sound of leaves rustling, or a flower blooming can all remind us of the beauty and interconnectedness of life.

So, next time you step outside, pause for a moment. Breathe in that fresh air, feel the sun's warmth on your skin, and open your heart to the synchronicities around you. Embrace the lessons nature's got for you, and let them guide you on your path. You'll find that as you cultivate this appreciation, you'll connect more deeply with the world—and with yourself.

In the end, it's all about realizing we're part of this amazing tapestry of life. Each thread, each moment, each synchronicity matters. So let's celebrate it, cherish it, and let it inspire us as we navigate our own journeys. You've got this!

Embrace the beauty of synchronicity in nature, and watch how it shifts your view on life.

# Chapter 13

**Practical Applications of Synchronicity**

Let's dive into the heart of synchronicity, shall we? This ain't just a fancy term to toss around at cocktail parties. Nope, it's a powerful tool you can use in your daily life. Think of synchronicity as the universe's way of nudging you, guiding you, and sometimes, giving you a good ol' shove in the right direction. So, how do you harness this energy? Well, let's break it down.

First off, applying the principles of synchronicity in your everyday life starts with awareness. You gotta be open to the signs. It's like tuning into a radio station—if you're not tuned in, you'll miss the music. Start by noticing patterns and coincidences in your life. Maybe you keep running into an old friend, or you hear a song that speaks to your current situation. Pay attention! These moments are the universe's way of saying, "Hey, look here!"

Next up, let's talk about techniques for recognizing and embracing those synchronous moments. One of the best ways to do this is through journaling. Grab a notebook and jot down those "aha" moments as they happen. Maybe you dream about a place and then see it in a movie the next day. Write it down! This practice not only helps you see the connections more clearly but also builds your intuition. Over time, you'll start to notice a rhythm in your life, like a dance that's been choreographed just for you.

Another great technique is mindfulness. You know that feeling when you're fully present in the moment? That's where the magic happens. Try meditation or simple breathing exercises. When your mind is calm, you're more likely to pick up on the subtle nudges from the universe. It's like clearing the static from that radio; suddenly, the music is crystal clear.

Now, let's create a personal synchronicity practice. This is where you take all those techniques and make 'em your own. Start with a daily ritual. Maybe it's a morning cup of coffee where you sit quietly and reflect on what you want to attract into your life. Or perhaps you take a walk in nature, allowing yourself to be open to whatever comes your way.

You can also set intentions. Intentions are like sending out a message in a bottle to the universe. What do you want to manifest? Write it down, say it out loud, or visualize it. This helps align your energy with what you're seeking.

And don't forget to celebrate those synchronous moments when they happen! Did you bump into someone who offered you a job lead? Did a chance encounter lead you to a new passion? Give yourself a pat on the back! Acknowledge the universe's hand in your life. This celebration reinforces your connection to synchronicity and keeps you in the flow.

You might also want to create a synchronicity buddy. Find someone who's interested in this journey too. Share your experiences, bounce ideas off each other, and keep each other accountable. It's like having a gym buddy, but for your spiritual growth. Together, you can explore the deeper meanings behind your experiences and learn from each other's insights.

Finally, don't forget to trust the process. Synchronicity isn't always about instant results. Sometimes, it's about planting seeds and allowing them to grow. You might not see the fruits of your labor right away, but that doesn't mean they aren't there. Stay patient and keep your heart open.

As you embark on this journey, remember that synchronicity is all about connection—between you and the universe, between you and others, and between your inner self and the world around you. Each synchronous moment is a reminder that you're not alone. You're part of a larger tapestry, woven together with threads of meaning, purpose, and timing.

So, go ahead! Embrace these principles, recognize those moments, and create a practice that resonates with you. You've got the tools. Now, it's time to put 'em to work. You're not just a passive observer in this life; you're an active participant. The universe is waiting for you to step into your power.

In the end, it's all about joy. The joy of discovery, the joy of connection, and the joy of knowing that you're right where you need to be. So, get out there, keep your eyes peeled, and watch as the magic of synchronicity unfolds before you. You've got this!

# Chapter 14

**The Future of Synchronicity**

As we edge into tomorrow, it's wild to think about how our grasp of synchronicity will change. The world's moving faster than a kid on a sugar high, and with all this chaos, our views are bound to shift. Picture future generations looking back at us, scratching their heads, trying to figure out our take on time and coincidences. Crazy, right?

Let's jump into this. Right now, synchronicity feels like a secret handshake between us and the universe. But as we dig deeper, it might just turn into something more than quirky fate. We're all part of this cosmic dance, where every move we make connects with others. Down the line, people might see synchronicity not just as random luck, but as a core part of existence.

Imagine a future where time isn't so rigid. We're stuck thinking linearly—past, present, future, like beads on a string. But what if future folks view time as a vast ocean? Waves of experiences crashing into each other? That'd flip everything upside down! They could realize that coincidences aren't just random; they're echoes from the past shaping the now and influencing the future.

In a world that's changing at lightning speed, synchronicity could be a lifeline. With tech racing ahead, we're hit with info overload. It's easy to feel adrift, like a boat lost at sea. But synchronicity could be that guiding star. Imagine future folks relying on these meaningful coincidences to steer their lives. They might say, "I met that person just when I needed to," or "I found that opportunity at the perfect time." Those moments would anchor them, reminding them they're not alone in this whirlwind.

Now, think about how younger generations might see time and coincidences. We often think of time as a straight line, but I bet they'll view it as a tapestry—threads weaving in and out, creating patterns that mirror their lives. They'll see coincidences not as flukes but as signs from the universe nudging them toward their paths. They might even create a whole new language around synchronicity, sharing tales of how chance meetings led to life-altering opportunities.

And let's not ignore tech's role in all this. With AI and virtual reality on the rise, who knows how synchronicity will show up? It could turn into a digital thing, where algorithms spot patterns in our lives, highlighting those serendipitous moments. Imagine getting a ping that says, "Hey, remember that book you were thinking about? Your buddy just posted about it!" That could change the game! Tech might help us notice connections we usually miss, making synchronicity more accessible.

But hey, there's a catch. We gotta be careful. As we lean on tech to make sense of our lives, we might lose touch with our gut feelings. Future generations will need to strike a balance—embracing the signs while trusting their instincts. It's like riding a bike; you gotta keep pedaling but stay aware of what's around you. They'll need to listen to that inner voice, the one saying, "Go this way," or "Turn left here."

Ultimately, the future of synchronicity will be a blend of old and new. It's a chance to redefine how we relate to time and coincidences. As we look ahead, let's remember that synchronicity isn't just about the big stuff; it's also in the little things—the smiles from strangers, unexpected calls, those gentle nudges reminding us we're all connected.

So, as you set out on your journey, keep your heart wide open. Embrace the possibilities synchronicity offers. Future generations will look back and say, "They opened the door for us to see this magic." And you, my friend, are part of that story. You're shaping how we think about time and connection, and that's pretty awesome.

Now, take a sec to picture your future. Imagine being surrounded by people and experiences that vibe with your soul. Feel that energy flowing through you, like a river carving its way through the land. That's synchronicity in action! You're not just watching from the sidelines; you're a key player in this cosmic dance.

As we wrap this up, let's hold onto that sense of wonder. The future's looking bright, and synchronicity's gonna be a big part of how we navigate it. So, keep your eyes peeled for those magical moments, my friend. They're out there, just waiting for you to find them. And remember, you've got this! Embrace the ride, and let synchronicity lead you to your destiny.

# Chapter 15

**Embracing Your Unique Journey**

Life's a crazy ride, right? One minute you're just going through the motions, and the next, everything clicks. It's like you're in some feel-good movie where everything aligns perfectly. That's synchronicity, folks. It's the universe giving you a little nudge, guiding you down your path. So, let's pause for a sec and think about your own moments of synchronicity. You know those times when everything just felt right? When the stars aligned, or you met someone who flipped your world upside down?

Take a breath and think back. Remember when you found that book that hit you right in the feels? Or when you ran into an old buddy just when you needed a pep talk? Those aren't just coincidences; they're the universe's way of saying, "Hey, you're not alone in this."

Personal growth is at the heart of these moments. Each synchronicity is like a stepping stone, pushing you to evolve and expand your view of the world. When you notice these happenings, you're not just sitting back; you're diving into life. You're saying, "I see you, universe! Let's grow!" And that's pretty awesome.

So, how do you tune into this awareness? Start a journal. Jot down those lucky moments as they pop up. You'll be shocked at how often they show up when you're actually looking. Reflect on what they mean. What are you learning? How are these experiences shaping your journey?

Now, let's chat about gratitude. Life gets busy, but taking a moment to appreciate how everything's connected can work wonders. Gratitude opens you up, helping you see the threads that tie us all together. It's like stepping back to see the whole picture, realizing that every person you meet and every challenge you face is part of a grand design by the universe.

When you embrace gratitude, you're not just focusing on the good stuff; you're also recognizing the lessons in the tough times. Maybe you hit a snag, but that led to a surprise opportunity. Or a relationship fizzled out, making room for something even better. Every experience, good or bad, is a chance to grow.

How do you build this gratitude habit? Start small. Each morning, write down three things you're thankful for. They can be as simple as that first sip of coffee or the sun peeking through your window. As you keep this up, you'll notice your perspective shifting. You'll see beauty in the everyday and recognize the synchronicities around you.

Now, let's tie it all together. Reflecting on synchronicity, embracing personal growth, and practicing gratitude aren't just separate tasks—they're all connected. Every time you spot a synchronous moment, you grow. And every time you grow, your gratitude for life deepens. It's a beautiful cycle that creates a rich tapestry of experiences that shape who you are.

And hey, remember, your journey is one of a kind. No one else has walked your path or faced your challenges. Embrace that! Celebrate it! You've got a voice that deserves to be heard, a story that needs telling.

As you move forward, keep your eyes peeled for those little signs from the universe. They're everywhere, just waiting for you to notice. Maybe it's a song that hits home or a stranger's smile that brightens your day. Those are the moments that remind you of your connection to something bigger.

Now, let's visualize your journey. Picture yourself strolling down a winding path, each step revealing new experiences, new people, and new synchronicities. Feel that excitement bubbling inside as you embrace the unknown. You're not just wandering; you're on a quest for growth, connection, and understanding.

And when you trip—because let's face it, you will—remember it's all part of the ride. Each stumble is a lesson, a chance to learn. Don't shy away from those moments; lean into them. Embrace the discomfort and trust it's leading you somewhere great.

As you navigate this journey, let gratitude be your compass. It'll guide you through the storms and help you savor the sunny days. Cultivating gratitude creates a mindset that attracts more synchronicity into your life. You'll start noticing the magic that's been there all along.

So, here's your challenge: Reflect on your synchronicity moments. Write them down, share them, and let them inspire you. Embrace the personal growth that comes from recognizing these experiences. And above all, cultivate gratitude for the beautiful, intricate web of life that connects us all.

You've got this! Your journey is unfolding just as it should, and the universe is rooting for you every step of the way. Keep your heart open, your mind curious, and your spirit adventurous. The best is yet to come, my friend. Now go out there and embrace your unique journey!

# Index

abilities, 12, 49, 53
Applications, 15, 98
appreciation, 15, 44, 57, 97, 98
artists, 13
Balancing, 14
beliefs, 14, 26, 28, 84
changing, 15, 37, 93, 105
cognitive, 5, 11, 26, 30
coincidence, 6, 10, 11, 19, 23, 27, 28, 30, 34, 57, 66, 70, 73, 74, 78, 88, 90
Concept, 11, 32

connection, 5, 12, 19, 20, 22, 38, 43, 45, 46, 49, 50, 53, 57, 58, 61, 65, 66, 68, 69, 72, 82, 88, 90, 91, 92, 95, 101, 102, 103, 108, 112, 113
creative, 13, 69, 70, 71, 72
Creativity, 13, 68, 69
cultivate, 12, 14, 15, 21, 45, 49, 53, 75, 77, 83, 91, 98, 114
cultures, 6, 12, 33, 34
cyclical, 5, 12, 32, 33, 34

daily, 6, 15, 49, 88, 99, 100
decision-making, 12, 51
deeper, 5, 6, 7, 15, 18, 20, 28, 29, 30, 35, 43, 44, 62, 64, 65, 95, 102, 104
Doubt, 13, 74
embracing, 15, 80, 85, 99, 107, 112
emotional, 6, 12, 44
events, 5, 6, 11, 18, 19, 26, 29, 30, 35, 60, 75, 85
evolving, 15
exercises, 6, 13, 70, 79, 100

experiences, 8, 10, 12, 14, 16, 18, 22, 23, 26, 29, 33, 34, 37, 38, 41, 42, 43, 44, 45, 46, 48, 50, 57, 59, 65, 70, 72, 74, 75, 76, 77, 79, 84, 89, 97, 102, 105, 108, 110, 112, 113, 114
Exploring, 11, 13
Fluid, 11, 32
friendships, 13, 56
Future, 15, 104, 107
generations, 16, 104, 106, 107
gratitude, 16, 44, 111, 112, 113, 114
growth, 16, 38, 43, 64, 65, 102, 110, 112, 113, 114
impact, 13, 56, 58
Impact, 14, 87
implications, 12, 20
importance, 14, 16, 58

individuals, 4, 12, 57, 58, 71, 90
innovators, 13, 69
inspiration, 7, 13, 18, 69, 70, 72, 73, 76
interconnectedness, 15, 16, 18, 44, 58, 82, 94, 96, 97
interpret, 14, 28
interpretation, 11
intersection, 11
Intuition, 12, 48, 51
intuitive, 12, 22, 49, 53
Journey, 16, 109
Jung, 5, 11, 17, 18, 19, 20
lessons, 9, 15, 58, 60, 98, 111
life, 6, 7, 9, 10, 12, 15, 16, 18, 20, 21, 22, 23, 26, 30, 33, 34, 35, 38, 40, 41, 42, 43, 44, 46, 47, 48, 49, 50, 51, 52, 53, 55, 57, 59, 60, 61, 64, 65, 66, 67, 70, 74, 75, 76, 77, 78, 80, 82, 83, 84, 88, 89, 90, 93, 94, 96, 97, 98, 99, 100, 101, 103, 106, 110, 112, 113, 114
life examples, 12
linear, 5, 12, 32, 34, 36
lives, 5, 6, 7, 8, 10, 13, 21, 29, 41, 42, 44, 45, 46, 48, 58, 63, 64, 83, 85, 87, 92, 94, 95, 105, 106, 107
Meaning, 11, 17
measure, 12, 34

mechanisms, 5, 11
messages, 13, 64, 84
mindfulness, 14, 34, 50, 53, 76, 79, 89, 91, 100
mindset, 6, 14, 34, 36, 75, 77, 113
modern, 14, 87
moments, 6, 8, 9, 10, 15, 18, 19, 22, 25, 28, 30, 35, 37, 40, 42, 45, 46, 47, 49, 53, 56, 57, 58, 60, 61, 62, 67, 70, 71, 72, 75, 76, 77, 78, 82, 83, 84, 85, 88, 89, 90, 92, 93, 95, 97, 99, 100, 101, 103, 105, 106, 108, 109, 110, 112, 113, 114
natural, 15, 37, 77, 93

Nature, 2, 5, 6, 9, 15, 37, 93
Observing, 15
open, 9, 13, 14, 23, 30, 31, 36, 37, 42, 45, 47, 59, 60, 63, 64, 66, 68, 69, 72, 75, 76, 77, 79, 80, 83, 87, 91, 92, 98, 99, 101, 102, 107, 114
Overcoming, 13, 74
partnerships, 13, 56, 57
perceive, 5, 12, 16
perception, 12, 14, 25, 26, 28, 30, 35, 87
Personal, 12, 40, 76, 110
phenomenon, 13, 29, 40, 55, 85
physics, 27
potential, 6, 15, 21, 23, 66

power, 14, 22, 38, 45, 50, 60, 76, 77, 81, 82, 83, 85, 92, 103
Practical, 6, 13, 15, 98
practice, 15, 45, 51, 63, 64, 76, 79, 90, 95, 100, 103
psychological, 11, 12, 28, 29
quantum, 8, 11, 25, 27, 29, 31
quantum physics, 8, 11, 25, 27, 29, 31
rapidly, 15
readers, 6, 16
recognizing, 15, 45, 47, 99, 111, 114
reflect, 4, 6, 16, 21, 23, 29, 30, 46, 60, 63, 89, 92, 95, 101
relationship between, 12, 14, 51
role, 9, 11, 14, 70, 76, 91, 94, 106
Science, 11, 25
significance, 13, 18, 28, 46
Signs, 13, 62
Skepticism, 13, 74
social media, 6, 14, 49, 88
Speculating, 15
spiritual, 6, 14, 81, 82, 83, 84, 85, 102
Spirituality, 14, 81
Stories, 12, 40
storytelling, 12, 41, 43
Strategies, 14
symbols, 9, 13, 63, 64, 65, 67, 78
Synchronicity, 2, 5, 6, 11, 12,

13, 14, 15, 17, 40, 45, 55, 56, 58, 68, 70, 81, 87, 93, 98, 102, 104
synchronous, 14, 15, 57, 59, 60, 75, 79, 83, 84, 89, 100, 101, 102, 112
Techniques, 15
Technology, 14, 87
Time, 2, 5, 6, 11, 32, 38
traditions, 14, 81
transformative, 10, 14, 17, 83, 85, 90

trust, 5, 7, 12, 36, 38, 47, 49, 51, 54, 58, 59, 60, 64, 72, 73, 95, 102, 113
Understanding, 12
Unique, 16, 109
Universe, 13, 62
world, 6, 8, 9, 15, 18, 22, 23, 28, 29, 31, 33, 34, 36, 37, 49, 60, 71, 75, 78, 79, 87, 88, 89, 91, 92, 93, 94, 96, 98, 102, 104, 105, 109, 110

www.ingramcontent.com/pod-product-compliance
Lightning Source LLC
Chambersburg PA
CBHW031438210526
45464CB00005B/2250